~A BINGO BOOK~

Animal Families Bingo Book

COMPLETE BINGO GAME IN A BOOK

Written By Rebecca Stark

PHOTO CREDITS:
"Lion Cub with Mother in the Serengeti," Tanzania, 2007
Cropped from http://www.flickr.com/photos/davidden/390536544/
Author: David Dennis

ISBN 978-0-87386-465-7

Educational Books 'n' Bingo

Printed in the U.S.A.

ANIMAL FAMILIES BINGO DIRECTIONS

INCLUDED:

List of Terms

Templates for Additional Terms and Clues

2 Clues per Term

30 Unique Bingo Cards

Markers

1. **Either cut apart the book or make copies of ALL the sheets. You might want to make an extra copy of the clue sheets to use for introduction and review. Keep the sheets in an envelope for easy reuse.**

2. Cut apart the call cards with terms and clues.

3. Pass out one bingo card per student. There are enough for a class of 30.

4. Pass out markers. You may cut apart the markers included in this book or use any other small items of your choice.

5. Decide whether or not you will require the entire card to be filled. Requiring the entire card to be filled provides a better review. However, if you have a short time to fill, you may prefer to have them do the just the border or some other format. Tell the class before you begin what is required.

6. There are 50 terms. Read the list before you begin. If there are any terms that have not been covered in class, you may want to read to the students the term and clues before you begin.

7. There is a blank space in the middle of each card. You can instruct the students to use it as a free space or you can write in answers to cover terms not included. Of course, in this case you would create your own clues. (Templates provided.)

8. Shuffle the cards and place them in a pile. Two or three clues are provided for each term. If you plan to play the game with the same group more than once, you might want to choose a different clue for each game. If not, you may choose to use more than one clue.

9. Be sure to keep the cards you have used for the present game in a separate pile. When a student calls, "Bingo," he or she will have to verify that the correct answers are on his or her card AND that the markers were placed in response to the proper questions. Pull out the cards that are on the student's card keeping them in the order they were used in the game. Read each clue as it was given and ask the student to identify the correct answer from his or her card.

10. If the student has the correct answers on the card AND has shown that they were marked in response to the *correct questions,* then that student is the winner and the game is over. If the student does not have the correct answers on the card OR he or she marked the answers in response to *the wrong questions,* then the game continues until there is a proper winner.

11. If you want to play again, reshuffle the cards and begin again.

Have fun!

TERMS

BEAVER	IMITATE
BUCK	JOEY
BULL	KID
CALF	KIT
CAT	LAMB
CATERPILLAR	LIONESS
COLONY	MARE
COLT	MIGRATE
COW	NEST
CUB	NOCTURNAL
DOE	PACK
DUCKLING	PENGUIN
EAGLET	PIG
EGG	PRAIRIE DOG
EWE	PRIDE
FAWN	RABBIT
FLOCK	ROOSTER
FOAL	PUP
FOWL	SOW
GAGGLE	STALLION
GOSLING	SWAN
HEN	TADPOLE
HERD	TROOP
HIBERNATE	WARREN
HONEYCOMB	WHELP

Additional Terms

Choose as many animal families terms as you would like and write them in the squares. Repeat each as desired.
Cut out the squares and randomly distribute them to the class.
Instruct the students to place their square on the center space of their card.

© **Barbara M. Peller.**

Clues for Additional Terms

Write three clues for each of your animal families terms.

_____ 1. 2. 3.	_____ 1. 2. 3.
_____ 1. 2. 3.	_____ 1. 2. 3.
_____ 1. 2. 3.	_____ 1. 2. 3.

BEAVER	**BUCK**
1. This animal is a rodent. It is known for building dams in rivers and streams. 2. The babies of this rodent are called kits. Its home is called a lodge.	1. A male deer or kangaroo. 2. When the female of a species is called a doe, then the male is called a ___.
BULL	**CALF**
1. This is the name for adult male cattle and also for a male elk and moose. 2. This is the name for adult male cattle and also for a male elephant, whale, and other animals.	1. A baby cow. 2. Although we usually think of a baby cow, this is also the name for a young camel, elephant, moose, whale, and other animals.
CAT	**CATERPILLAR**
1. A young one is called a kitten. 2. Like lions and tigers, the ___ is a feline.	1. It is the larva stage of a butterfly or moth. 2. Someday it will turn into a butterfly.
COLONY	**COLT**
1. Ants are social insects and live in a ___. 2. An ant ___ may have millions of ants.	1. A young male horse is a ___. 2. A young male pony is a ___.
COW	**CUB**
1. If the male of a species is called a bull, then the female is called a ___. 2. The adult female of cattle and several other animals, such as elephants and whales, is called this.	1. A young bear, wolf, or lion, for example. A synonym is "whelp." 2. The young of certain carnivorous (meat-eating) animals, such as a bear, fox, or tiger. A synonym is "whelp."

Animal Families Bingo

DOE	**DUCKLING**
1. A female deer.	1. A baby duck.
2. When the male of a species is called a buck, then the female is called a ___ .	2. Like a gosling, this baby bird has a covering of soft, fluffy feathers called down.
EAGLET	**EGG**
1. A baby eagle.	1. Mammals give birth to live young. A reptile or bird baby hatches from an ___ .
2. Its nest, like that of other birds of prey, is called an aerie.	2. A reptile or bird ___ has a protective shell.
EWE	**FAWN**
1. A female sheep.	1. A young deer.
2. Her baby is called a lamb.	2. Its mother is a doe and its father is a buck.
FLOCK	**FOAL**
1. Sheep and birds travel in a ___ .	1. A young horse of either sex, especially under a year old, is a ___ .
2. There is a saying: "Birds of a feather, ___ together."	2. A young male pony of either sex, especially under a year old, is a ___ .
FOWL	**GAGGLE**
1. Chickens and turkeys are members of this group of birds.	1. A flock of geese when not in flight is often called this.
2. The females of this group are called hens.	2. When in flight, a flock of geese is often called a skein instead of a ___ .

Animal Families Bingo

GOSLING 1. A baby goose. 2. Its father is called a gander.	**HEN** 1. An adult female chicken. 2. Most of the eggs that people eat come from one.
HERD 1. A ___ is a large group of cattle or other animal feeding and traveling together. 2. Elephants travel in a ___. So do caribou and deer.	**HIBERNATE** 1. To pass the winter in a resting state. 2. In cold regions squirrels ___ during the winter.
HONEYCOMB 1. Honey bees build one to hold honey and larvae. 2. A structure of small 6-sided cells made by bees	**IMITATE** 1. To copy. 2. Babies ___ their parents to learn. They use their parents as an example.
JOEY 1. A baby kangaroo. 2. This baby spends some time in its mother's pouch.	**KID** 1. A young goat. 2. Its mother is a doe, but sometimes she is called a nanny. Its father is sometimes called a billy.
KIT 1. Sometimes a young fox is called a cub, but it is also called a ___. 2. This name for a baby fox sounds almost like a baby cat.	**LAMB** 1. A baby sheep. 2. Its mother is a ewe. Its father is a ram.

Animal Families Bingo

© **Barbara M. Peller.**

LIONESS

1. A female lion is a ___.

2. The female lion, or ___, usually hunts with female relatives.

MARE

1. An adult female horse is a ___.

2. Until she is four years old, a ___ is called a filly.

MIGRATE

1. To go seasonally from one region or climate to another is to ___.

2. Many birds ___ to warmer places in the winter.

NEST

1. An eagle's ___ is called an aerie.

2. A pigeon's ___ is made of small twigs. Adult pigeons build their ___ in hard-to-reach places.

NOCTURNAL

1. A ___ animal is active during the night.

2. Many owls are ___. So are bats. They are active at night.

PACK

1. Wolves and other wild dogs travel in a ___.

2. A group of predatory animals of the same kind is often called this.

PENGUIN

1. This bird does not fly.

2. ___ parents keep their chicks well fed and protected in areas called rookeries.

PIG

1. A baby one is called a piglet.

2. A male adult is called a boar. A female adult is called a sow.

PRAIRIE DOG

1. This animal is a rodent although its name might make you think differently.

2. This rabbit-sized rodent lives in a community of underground burrows with its family group.

PRIDE

1. Lions live in a large social group called a pride. They are the only cats to do so.

2. A lion ___ is made up of 3 to 30 lions.

Animal Families Bingo

RABBIT 1. A baby ___ is called a bunny. 2. It lives in a burrowed area called a warren.	**ROOSTER** 1. An adult male chicken. 2. It says, "Cock-a doodle doo."
PUP 1. Although it is sometimes called a cub, a young seal is usually called a ___. 2. This name for a young seal is the same as the shortened form of the name for a young dog.	**SOW** 1. An adult female pig. 2. Its baby is called a piglet.
STALLION 1. An adult male horse. 2. A ___ is more muscular than the female, or mare.	**SWAN** 1. A young one is called a cygnet. 2. A female ___ is called a pen and a male is called a cob.
TADPOLE 1. The aquatic larval stage of an amphibian. 2. Every frog was once this.	**TROOP** 1. Baboons travel in a ___. 2. Gorillas travel in a ___. It is sometimes called a band.
WARREN 1. Rabbits live in a burrowed area called a ___. 2. A structure where rabbits are kept.	**WHELP** 1. A young bear, wolf, or lion, for example. A synonym is "cub." 2. The young of certain carnivorous (meat-eating) animals, especially the dog. A synonym is "cub."

Animal Families Bingo

Animal Families Bingo

Herd	Fowl	Pup	Warren	Swan
Colt	Beaver	Troop	Lamb	Hen
Pride	Penguin		Honeycomb	Lioness
Whelp	Nest	Flock	Rooster	Hibernate
Nocturnal	Duckling	Cow	Fawn	Foal

Animal Families Bingo: Card No. 1

Animal Families Bingo

Animal Families Bingo

Whelp	Pup	Joey	Mare	Nocturnal
Hibernate	Lamb	Buck	Nest	Prairie Dog
Pack	Duckling		Cub	Flock
Ewe	Rabbit	Penguin	Imitate	Hen
Foal	Troop	Cow	Colt	Fawn

Animal Families Bingo

Whelp	Flock	Lamb	Rooster	Pride
Duckling	Beaver	Cat	Fowl	Gosling
Nest	Troop		Prairie Dog	Bull
Penguin	Pack	Nocturnal	Ewe	Joey
Fawn	Colt	Cow	Imitate	Pup

Animal Families Bingo

Penguin	Prairie Dog	Nocturnal	Colt	Sow
Kid	Buck	Fowl	Mare	Pride
Honeycomb	Ewe		Swan	Warren
Flock	Pig	Troop	Cow	Cat
Gaggle	Foal	Kit	Fawn	Lioness

Animal Families Bingo

Foal	Swan	Nest	Buck	Colt
Kid	Flock	Cat	Cub	Beaver
Pup	Lioness		Doe	Eaglet
Hen	Prairie Dog	Herd	Imitate	Gaggle
Lamb	Cow	Pig	Penguin	Honeycomb

Animal Families Bingo: Card No. 5

Animal Families Bingo

Bull	Prairie Dog	Joey	Pup	Lioness
Rooster	Nest	Gaggle	Fowl	Pride
Mare	Caterpillar		Buck	Cub
Cow	Nocturnal	Imitate	Kit	Sow
Hibernate	Flock	Herd	Honeycomb	Pig

Animal Families Bingo: Card No. 6

Animal Families Bingo

Herd	Prairie Dog	Eaglet	Doe	Lamb
Hibernate	Nocturnal	Duckling	Beaver	Kid
Joey	Warren		Cub	Calf
Penguin	Ewe	Cat	Whelp	Pack
Cow	Colt	Imitate	Kit	Bull

Animal Families Bingo

Honeycomb	Prairie Dog	Colony	Rooster	Calf
Kid	Pup	Mare	Lioness	Buck
Pride	Migrate		Sow	Swan
Fawn	Penguin	Whelp	Gaggle	Ewe
Troop	Cow	Kit	Nest	Hibernate

Animal Families Bingo

Cub	Lamb	Duckling	Pride	Lioness
Gaggle	Pup	Honeycomb	Nest	Sow
Gosling	Herd		Beaver	Colony
Caterpillar	Foal	Nocturnal	Doe	Eaglet
Ewe	Imitate	Cat	Whelp	Swan

Animal Families Bingo: Card No. 9

Animal Families Bingo

Whelp	Rooster	Buck	Mare	Pig
Lioness	Calf	Fowl	Beaver	Sow
Migrate	Prairie Dog		Warren	Pack
Nocturnal	Hen	Gaggle	Imitate	Gosling
Cat	Hibernate	Joey	Foal	Honeycomb

Animal Families Bingo

Bull	Prairie Dog	Nest	Gaggle	Hibernate
Colony	Gosling	Doe	Cub	Fowl
Kid	Pup		Joey	Duckling
Cat	Pride	Imitate	Colt	Whelp
Caterpillar	Cow	Herd	Kit	Lamb

Animal Families Bingo

Lamb	Swan	Gosling	Rooster	Cub
Duckling	Troop	Pup	Kit	Beaver
Herd	Eaglet		Lioness	Mare
Cow	Ewe	Sow	Whelp	Kid
Prairie Dog	Colony	Migrate	Caterpillar	Calf

Animal Families Bingo

Caterpillar	Swan	Bull	Gosling	Lioness
Pup	Colony	Prairie Dog	Cub	Pack
Rooster	Buck		Duckling	Eaglet
Honeycomb	Imitate	Calf	Migrate	Whelp
Cow	Hen	Kit	Herd	Doe

Animal Families Bingo: Card No.13

© Barbara M. Peller.

Animal Families Bingo

Colt	Pup	Nest	Cub	Caterpillar
Calf	Herd	Gosling	Beaver	Prairie Dog
Gaggle	Warren		Joey	Cat
Hen	Imitate	Migrate	Buck	Bull
Cow	Mare	Pack	Hibernate	Honeycomb

Animal Families Bingo: Card No. 14

Animal Families Bingo

Doe	Cub	Nest	Lamb	Rooster
Bull	Joey	Fowl	Pup	Gaggle
Lioness	Herd		Pride	Sow
Cow	Gosling	Colony	Imitate	Caterpillar
Hibernate	Ewe	Kit	Pig	Duckling

Animal Families Bingo: Card No. 15

Animal Families Bingo

Buck	Gosling	Colony	Pig	Rabbit
Mare	Pack	Eaglet	Kid	Warren
Caterpillar	Swan		Lioness	Duckling
Penguin	Calf	Cow	Doe	Whelp
Gaggle	Tadpole	Kit	Ewe	Prairie Dog

© Barbara M. Peller.

Animal Families Bingo

Cat	Stallion	Egg	Gosling	Colt
Doe	Gaggle	Imitate	Warren	Eaglet
Cub	Honeycomb		Tadpole	Colony
Foal	Hibernate	Whelp	Nest	Pack
Nocturnal	Caterpillar	Lamb	Rooster	Swan

Animal Families Bingo

Pig	Migrate	Calf	Gaggle	Mare
Prairie Dog	Cat	Nocturnal	Lioness	Caterpillar
Cub	Pack		Egg	Sow
Foal	Fowl	Imitate	Whelp	Joey
Tadpole	Gosling	Nest	Stallion	Bull

© **Barbara M. Peller.**

Animal Families Bingo

Lioness	Bull	Gosling	Colony	Migrate
Doe	Rooster	Sow	Lamb	Warren
Stallion	Colt		Beaver	Pig
Joey	Tadpole	Nocturnal	Ewe	Egg
Pride	Rabbit	Hibernate	Honeycomb	Kit

Animal Families Bingo: Card No. 19

Animal Families Bingo

Migrate	Stallion	Rooster	Gosling	Beaver
Buck	Duckling	Kid	Nocturnal	Mare
Swan	Eaglet		Penguin	Fowl
Foal	Troop	Fawn	Ewe	Tadpole
Flock	Honeycomb	Rabbit	Whelp	Egg

Animal Families Bingo: Card No. 20

Animal Families Bingo

Doe	Bull	Kid	Gosling	Hen
Swan	Egg	Calf	Colony	Herd
Pack	Hibernate		Stallion	Nest
Nocturnal	Lamb	Tadpole	Foal	Honeycomb
Penguin	Rabbit	Kit	Cat	Ewe

Animal Families Bingo: Card No. 21

Animal Families Bingo

Pride	Joey	Egg	Pup	Caterpillar
Mare	Rooster	Pig	Colony	Beaver
Calf	Warren		Herd	Eaglet
Tadpole	Foal	Ewe	Fowl	Colt
Rabbit	Cat	Stallion	Pack	Kid

Animal Families Bingo

Buck	Stallion	Lamb	Pup	Kit
Bull	Migrate	Hibernate	Doe	Fowl
Joey	Caterpillar		Fawn	Herd
Pack	Rabbit	Tadpole	Cat	Ewe
Hen	Troop	Honeycomb	Nocturnal	Egg

Animal Families Bingo: Card No. 23

Animal Families Bingo

Buck	Migrate	Colt	Stallion	Colony
Lioness	Kit	Kid	Mare	Herd
Eaglet	Pig		Caterpillar	Pack
Hen	Fawn	Tadpole	Cat	Swan
Flock	Penguin	Rabbit	Rooster	Troop

Animal Families Bingo

Penguin	Kid	Stallion	Nest	Egg
Fowl	Hen	Doe	Calf	Beaver
Swan	Colony		Fawn	Tadpole
Pig	Foal	Troop	Rabbit	Warren
Kit	Colt	Eaglet	Gaggle	Flock

Animal Families Bingo

Egg	Stallion	Joey	Mare	Pig
Nocturnal	Rooster	Colony	Migrate	Buck
Hen	Fawn		Warren	Penguin
Cat	Pup	Foal	Rabbit	Tadpole
Eaglet	Gaggle	Nest	Troop	Flock

Animal Families Bingo: Card No. 26

Animal Families Bingo

Joey	Calf	Stallion	Migrate	Duckling
Hen	Fawn	Doe	Tadpole	Beaver
Imitate	Troop		Rabbit	Penguin
Pig	Bull	Kid	Flock	Fowl
Caterpillar	Warren	Egg	Pride	Eaglet

Animal Families Bingo: Card No. 27

Animal Families Bingo

Lioness	Migrate	Pig	Stallion	Calf
Duckling	Egg	Fawn	Mare	Warren
Troop	Pack		Eaglet	Nocturnal
Whelp	Pride	Hibernate	Rabbit	Tadpole
Pup	Cub	Caterpillar	Flock	Hen

Animal Families Bingo

Egg	Migrate	Pig	Doe	Cub
Hen	Nocturnal	Kid	Eaglet	Pride
Swan	Fawn		Beaver	Stallion
Duckling	Foal	Pup	Rabbit	Tadpole
Buck	Colony	Flock	Bull	Troop

Animal Families Bingo: Card No. 29

Animal Families Bingo

Colt	Stallion	Mare	Cub	Tadpole
Fowl	Migrate	Joey	Warren	Beaver
Hen	Caterpillar		Eaglet	Kid
Flock	Bull	Pup	Rabbit	Fawn
Foal	Lamb	Troop	Egg	Pig

Animal Families Bingo: Card No. 30